Oxford Read and Discover

2

D1330373

# Plastic

Louise Spilsbury

**Contents**

| Introduction | 3 |
| 1 What Is Plastic? | 4 |
| 2 Making Plastic | 6 |
| 3 Bottles and Boxes | 8 |
| 4 At Home | 10 |
| 5 Outside | 12 |
| 6 Plastic Waste | 14 |
| 7 Recycling Plastic | 16 |
| 8 Plastic Today | 18 |
| Activities | 20 |
| Projects | 36 |
| Picture Dictionary | 38 |
| About *Read and Discover* | 40 |

**OXFORD**

UNIVERSITY PRESS

# OXFORD
UNIVERSITY PRESS

Great Clarendon Street, Oxford, OX2 6DP, United Kingdom

Oxford University Press is a department of the University of Oxford. It furthers the University's objective of excellence in research, scholarship, and education by publishing worldwide. Oxford is a registered trade mark of Oxford University Press in the UK and in certain other countries

ISBN: 978 0 19 464688 8

An Audio Pack containing this book and an Audio download is also available, ISBN 978 0 19 402167 8

This book is also available as an e-Book, ISBN 978 0 19 410868 3.

An accompanying Activity Book is also available, ISBN 978 0 19 464678 9

Printed in China

This book is printed on paper from certified and well-managed sources.

ACKNOWLEDGEMENTS

*Illustrations by*: Kelly Kennedy pp.7, 8, 11, 16, 17; Alan Rowe pp.20, 21, 23, 24, 25, 26, 28, 30, 32, 33, 34, 38, 39.

*The Publishers would also like to thank the following for their kind permission to reproduce photographs and other copyright material*: Alamy pp.8 (Wrappers), 14 (fire/Tim Gainey), 15 (turtle/FLPA), 19 (cups/Tim Gainey); Corbis p.6 (pellets/Bob Krist); Getty Images pp.11 (sitting room/Johner Images), 13 (Mountain Biking UK Magazine), 14 (rubbish/Brent Stirton), 17 (Don Klumpp/Photographer's Choice), 18 (Johan Ordonez/Stringer/AFP), 19 (corn field/Bernard Jaubert/Workbook Stock); Oxford University Press pp.3 (chair, computer, phone, ball), 4 (pens, cd, sandals, bottle, cutlery, chopsticks), 5 (watch, skateboard), 6 (oil pump), 9 (plastic boxes, sandwich), 10, 12 (kite, kayak); Science Photo Library pp.11 (blue fibres/orange fibres/Susumu Nishinaga), 19 (bags/Environmental Images/UIG).

# Introduction

We use plastic every day. There's plastic in our homes, schools, and offices. There's plastic in our toys, computers, and telephones. Is there plastic in your chair, too?

What plastic things can you see here? What plastic things do you use?

Now read and discover more about plastic!

# What Is Plastic?

Plastic is a material. We use materials to make things. We use plastic to make many things in many different shapes. We use plastic to make pens, bottles, CDs, and shoes. We make plastic knives, forks, spoons, and chopsticks, too!

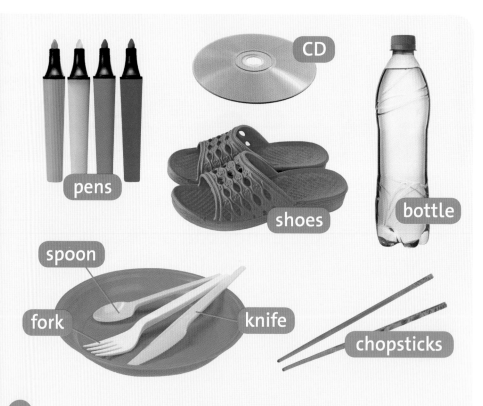

pens

CD

shoes

bottle

spoon

fork

knife

chopsticks

We make many different types of plastic. Some types of plastic are soft. A plastic watch strap is soft so we can put it on our wrist.

Some types of plastic are hard and strong. A plastic skateboard is hard and strong so we can stand on it.

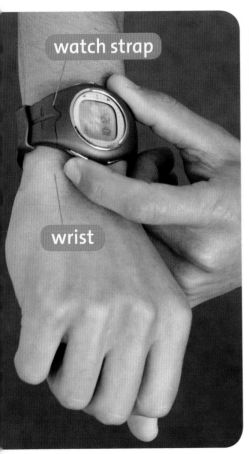

watch strap

wrist

skateboard

→ Go to pages 20–21 for activities.

# 2 Making Plastic

Getting Oil

We can make plastic from oil. We get oil from under the ground.

We can get different types of liquid from oil. We use some of the liquids to make small plastic pieces. We use the plastic pieces to make plastic things.

Plastic Pieces

In a factory, we make plastic pieces hot. Hot plastic is very soft. Then we can make it into different shapes.

How do we make plastic spoons? We make plastic pieces hot. We put the hot plastic in a mold. The mold has spoon shapes. The plastic gets cold and hard in the mold. Then we have lots of spoons!

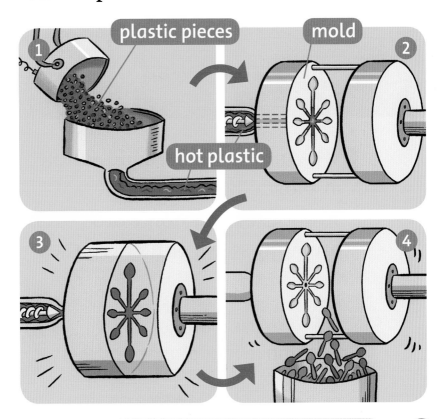

plastic pieces   mold

hot plastic

→ Go to pages 22–23 for activities.

Plastic Bottles

We can buy juice, soda, and water in plastic bottles. Plastic bottles are strong, so they don't break. They are light so we can carry them.

We use 1 liter of oil to make four plastic bottles!

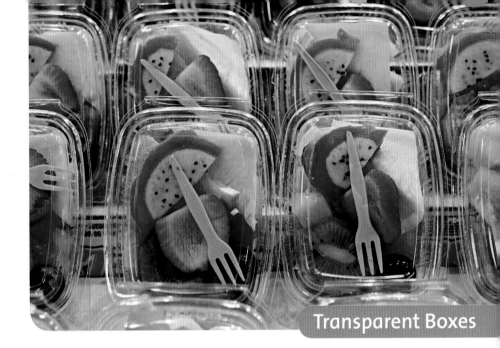

Transparent Boxes

Many plastic boxes are transparent. They are transparent so we can see what's in the boxes.

We can buy lots of food in plastic wrap. We can buy sandwiches, salad, and cake in plastic wrap. The food is in plastic wrap to stop it going bad.

Food in Plastic Wrap

→ Go to pages 24–25 for activities.

# 4 At Home

There's lots of plastic at home. We use plastic to make many doors, window frames, and roofs. Plastic is strong, and it stops rain getting into our homes. Plastic pipes take rain water from our roofs.

roof

pipe

window frame

door

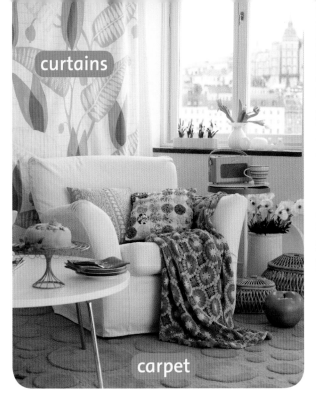

curtains

carpet

fibers

We use some plastic to make fibers. We can make plastic fibers in different colors. We use plastic fibers to make many carpets and curtains. We make lots of clothes from plastic fibers, too.

 Discover!

wall

We can make paint from plastic, too! We use it on the walls at home.

paint

→ Go to pages 26–27 for activities.

# 5 Outside

**Flying a Kite**

We use many plastic things outside. We make plastic toys in many shapes and colors. We play soccer with plastic balls. We hit balls with plastic bats. We fly plastic kites. We use plastic kayaks. What plastic things do you use outside?

**A Kayak**

# Oxford Read and Discover

## Audio Download

Level 2

**Plastic**

## To download the audio for this title

1  Go to www.oup.com/elt/download

2  Use this code and your email address

Your download code

ISBN 978-0-19-402168-5

9 780194 021685

helmet

goggles

pad

Plastic helps us to be safe when we are outside. Plastic goggles protect our eyes. Plastic pads protect our elbows and knees. Plastic helmets are light and very strong. They don't break when we fall. They protect our head. Do you wear a plastic helmet when you ride a bicycle?

→ Go to pages 28–29 for activities.

# 6 Plastic Waste

Every day, we throw away many plastic things. This makes plastic waste. Today, there's too much plastic waste.

Burning Plastic Waste

Some people burn plastic waste. This makes pollution. Pollution is very bad for us.

Some plastic waste goes in the ocean. Many animals live in the ocean. The plastic waste is very bad for the animals.

Some plastic things break into small pieces. Animals eat the pieces and they get sick. Some animals eat plastic bags and they get sick. Some plastic waste hurts animals, too.

Plastic Waste Hurts Animals

Go to pages 30–31 for activities.

# Recycling Plastic

We can recycle old plastic to make new things. This is how we recycle plastic bottles to make T-shirts.

1. We take plastic bottles to a recycling center.
2. We take the bottles to a factory.
3. We break the bottles into small plastic pieces.
4. We use the plastic pieces to make fibers.
5. We use the fibers to make T-shirts.

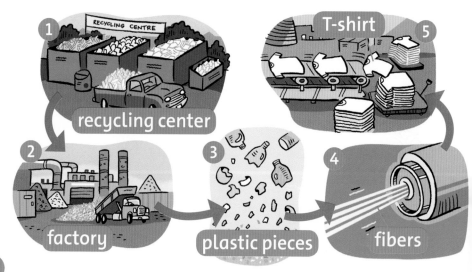

recycling center

factory

plastic pieces

fibers

T-shirt

A Playground

We recycle old plastic to make lots of new things. We recycle plastic bottles to make T-shirts, kayaks, chairs, carpets, and playgrounds. We recycle plastic bags and plastic bottles to make new bags and bottles, too!

Discover!

We use ten plastic bottles to make one T-shirt!

→ Go to pages 32–33 for activities.

# 8 Plastic Today

We use lots of oil to make plastic. We can't make more oil. We use lots of plastic every day, and there's too much plastic waste.

You can help. Don't use lots of plastic bags. Don't throw away plastic bottles. Recycle plastic bottles. Use plastic bottles to make new things, too.

There are plastic bottles in this boat! Can you see them?

A Boat

plants

cups

bags

100% Biodegradable

ecology center

PLEASE COMPOST THIS BAG

Today, we can make plastic from plants, too. We use this plastic to make cups and bags. We don't use oil to make this plastic. We can grow more plants to make more plastic!

→ Go to pages 34–35 for activities.

# 1 What Is Plastic?

← Read pages 4–5.

**1** Circle the correct words.

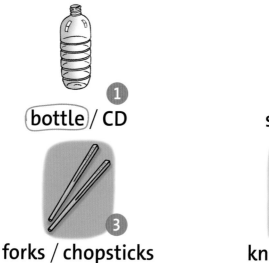

**1**
(bottle) / CD

**2**
shoe / pen

**3**
forks / chopsticks

**4**
knife / spoon

**2** Complete the sentences.

plastic   bottles   shapes   ~~material~~

1 Plastic is a __material__ .

2 We use plastic to make many things in
many different _____ .

3 We use plastic to make pens and _____ .

4 We use _____ to make knives and forks.

## 3 Find and write the words.

| s | o | f | t | b | o | n | z | o |
|---|---|---|---|---|---|---|---|---|
| x | q | f | t | i | h | a | r | d |
| z | r | u | q | p | c | n | l | x |
| p | l | a | s | t | i | c | e | k |
| b | d | a | g | c | s | r | t | w |
| s | t | r | o | n | g | a | l | s |

1 __plastic__

2  s_____    3  h_____    4  s_____

## 4 Write *true* or *false*.

1  We make one type of plastic.  __false__

2  Some types of plastic are soft.  _____

3  Some types of plastic are hard
   and strong.  _____

4  Plastic watch straps are hard
   and strong.  _____

5  Plastic skateboards are soft.  _____

6  We can stand on a plastic skateboard.  _____

## 2 Making Plastic

← Read pages 6–7.

**1 Find and write the words.**

plasticoilgroundliquidpiecesfactory

1 __plastic__     3 _____     5 _____

2 _____     4 _____     6 _____

**2 Match. Then write the sentences.**

We can make plastic — to make plastic things.

We get oil from — under the ground.

We can get — different types of liquid from oil.

We use plastic pieces — from oil.

1 __We can make plastic from oil.__

2 _____

3 _____

4 _____

## 3 Complete the sentences.

1 In a _____ , we make plastic pieces hot.

2 Hot plastic is very _____ .
We can make it into different _____ .

3 To make plastic spoons, we put hot plastic in a _____ . The mold has _____ shapes.

4 The plastic gets _____ and _____ in the mold.

5 Then we have lots of _____ .

# (3) Bottles and Boxes

← Read pages 8–9.

**1 Write the words.**

light   water
juice   bottle

1 _____      3 _____

2 _____      4 _____

**2 Answer the questions.**

1 What can we buy in plastic bottles?

   _We can buy juice, soda, and water in_
   _plastic bottles._

2 What can we buy in plastic wrap?

   _____

3 How many liters of oil do we use to make
   four plastic bottles?

   _____

# 3 Complete the puzzle. Then write the secret word.

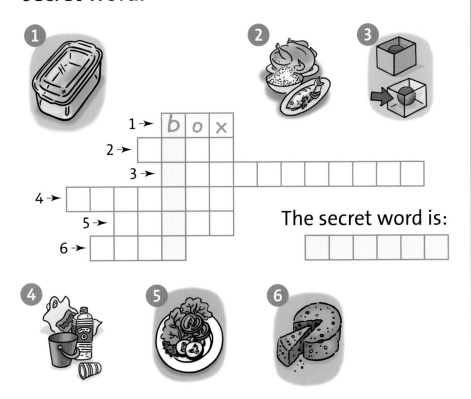

1 → | b | o | x |

2 →

3 →

4 →

5 →

6 →

The secret word is:

# 4 Write *true* or *false*.

1 No plastic boxes are transparent. _____

2 Plastic boxes are transparent
so we can't see what's in the boxes. _____

3 We can buy lots of food in plastic
wrap. _____

4 Food is in plastic wrap so it goes bad. _____

# ④ At Home

← Read pages 10–11.

## 1 Circle the correct words.

spoon / **door**

roof / skateboard

box / pipe

window frame / bottle

## 2 Order the words.

1 lots / at home. / of plastic / There's

   <u>There's lots of plastic at home.</u>

2 to make / We use / many doors. / plastic

   _____

3 stops / Plastic / getting into / rain / our homes.

   _____

## 3 Complete the sentences.

plastic   colors   curtains   make   fibers

1 We use some plastic to make _____ .

2 We can make plastic fibers in different
_____ .

3 We use plastic fibers to make many carpets
and _____ .

4 We make lots of our clothes from
_____ fibers.

5 We can _____ paint from plastic.

## 4 Answer the questions.

1 What do we use plastic fibers to make?

_____

2 What can we make paint from?

_____

3 Where do we use paint at home?

_____

4 Do you use paint at home?

_____

# 5 Outside

← Read pages 12–13.

## 1 Complete the puzzle.

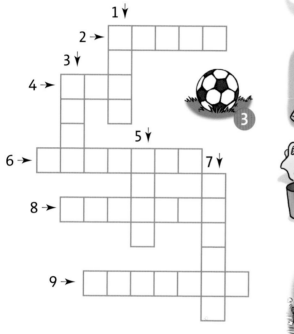

## 2 Write *true* or *false*.

1 We use many plastic things outside. _____

2 We play soccer with plastic spoons. _____

3 We hit balls with plastic kayaks. _____

4 We fly plastic kites. _____

5 We fly plastic kayaks. _____

## 3 Match. Then write the sentences.

| | |
|---|---|
| Plastic helps us | protect our eyes. |
| Plastic goggles | to be safe when we are outside. |
| Plastic pads | our head. |
| Plastic helmets are | when we fall. |
| Plastic helmets don't break | light and very strong. |
| Plastic helmets protect | protect our elbows and knees. |

1 _____

2 _____

3 _____

4 _____

5 _____

6 _____

## 4 Find and write the words.

padselbowsheadkneeshelmetgoggles

1 _____    3 _____    5 _____

2 _____    4 _____    6 _____

# 6 Plastic Waste

← Read pages 14–15.

## 1 Complete the sentences.

1 Every day, we _____ many _____ things.

2 Today, there's too much plastic _____ .

3 Some people _____ plastic waste and this makes pollution.

4 _____ is very bad for us.

## 2 Find and write the words.

oceananimalswasteburnpollutionbad

1 _____   3 _____   5 _____

2 _____   4 _____   6 _____

## 3 Order the words.

1 ocean. / Some / goes in the / plastic waste

_____

2 live in / Many / the ocean. / animals

_____

3 for the animals. / The / is very bad /
plastic waste

_____

## 4 Match. Then write the sentences.

| | |
|---|---|
| Some plastic things | hurts animals. |
| Animals eat plastic pieces | bags and they get sick. |
| Some animals eat plastic | and they get sick. |
| Some plastic waste | break into small pieces. |

1 _____

2 _____

3 _____

4 _____

# 7 Recycling Plastic

← Read pages 16–17.

## 1 Find and write the words.

| r | e | c | y | c | l | e | i |
|---|---|---|---|---|---|---|---|
| x | q | f | t | i | f | i | b |
| y | i | h | u | y | c | s | h |
| r | e | f | i | b | e | r | s |
| b | d | a | p | l | a | s | e |
| p | i | e | c | e | s | a | s |
| a | f | a | c | t | o | r | y |

1 r _____   2 f _____

3 f _____   4 p _____

## 2 Number the sentences in order.

☐ We break the bottles into small plastic pieces.

☐ We use the plastic pieces to make fibers.

☐ We take the bottles to a factory.

☐ We use the fibers to make T-shirts.

☐ 1 We take plastic bottles to a recycling center.

**3** **Write the words. Then match.**

1 y a k a k

<u>*kayak*</u>

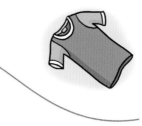

2 c r a i h

_____

3 i r t T- s h

_____

4 n d P g r l a o y u

_____

**4** **Write *true* or *false*.**

1 We recycle old plastic to make lots of old things. _____

2 We recycle plastic bottles to make T-shirts, kayaks, chairs, carpets, and playgrounds. _____

3 We recycle plastic bags and plastic bottles to make old bags and bottles. _____

4 We use two old plastic bottles to make one new T-shirt. _____

33

# 8 Plastic Today

← Read pages 18–19.

**1 Write the words.**

plastic bag    oil
plastic cup    plants

1 _____     3 _____

2 _____     4 _____

**2 Match. Then write the sentences.**

We use lots of oil          more oil.
We can't make              to make plastic.
We use plastic             plastic waste.
There's too much           every day.

1 _____

2 _____

3 _____

4 _____

**3** Complete the sentences.

plants    plastic    Recycle    bottles

1  Don't use lots of _____ bags.

2  Don't throw away plastic _____ .

3  _____ bottles.

4  We can make plastic from _____ .

**4** Order the letters and write the words.
Then write the secret word.

1  t c r n e e

2  n w e

3  u p s c

4  y t o a d

5  c p l t i a s

6  s b t l o t e

7  k e m a

1➤ | c | e | n | t | e | r |

2➤

3➤

4➤

5➤

6➤

7➤

The secret word is:

# A Plastic Poster

1. What plastic things are in your bedroom, kitchen, and classroom? Write the words.

| Bedroom | Kitchen | Classroom |
|---------|---------|-----------|
|         |         |           |
|         |         |           |
|         |         |           |
|         |         |           |
|         |         |           |
|         |         |           |
|         |         |           |
|         |         |           |
|         |         |           |

2. Make a poster. Find or draw pictures of the plastic things. Write about them.

3. Display your poster.

# A Plastic Diary

**1** Write the plastic things that you use in five days.

| Days | Plastic Things |
|------|----------------|
| Monday | |
| Tuesday | |
| Wednesday | |
| Thursday | |
| Friday | |

**2** Count the plastic things. Draw a graph.

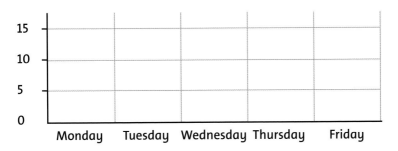

# Picture Dictionary

 break

 burn

 buy

 carry

 clothes

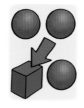 different

 factory

 fall

 food

 ground

 grow

 hard

 hurt

 light

 liquid

 materials

 ocean

 oil

 pieces

 plants

 plastic

 pollution

 protect

 recycle

 recycling center

 safe

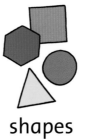 shapes

 soft

 strong

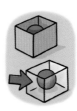 throw away

 transparent

waste

# Oxford Read and Discover

Series Editor: Hazel Geatches • CLIL Adviser: John Clegg

**Oxford Read and Discover** graded readers are at six levels, for students from age 6 and older. They cover many topics within three subject areas, and support English across the curriculum, or Content and Language Integrated Learning (CLIL).

Available for each reader:
- Audio Pack
- Activity Book

Available for selected readers:
- e-Books

Teaching notes & CLIL guidance: **www.oup.com/elt/teacher/readanddiscover**

| Subject Area / Level | The World of Science & Technology | The Natural World | The World of Arts & Social Studies |
|---|---|---|---|
| **1** — 300 headwords | • Eyes<br>• Fruit<br>• Trees<br>• Wheels | • At the Beach<br>• In the Sky<br>• Wild Cats<br>• Young Animals | • Art<br>• Schools |
| **2** — 450 headwords | • Electricity<br>• Plastic<br>• Sunny and Rainy<br>• Your Body | • Camouflage<br>• Earth<br>• Farms<br>• In the Mountains | • Cities<br>• Jobs |
| **3** — 600 headwords | • How We Make Products<br>• Sound and Music<br>• Super Structures<br>• Your Five Senses | • Amazing Minibeasts<br>• Animals in the Air<br>• Life in Rainforests<br>• Wonderful Water | • Festivals Around the World<br>• Free Time Around the World |
| **4** — 750 headwords | • All About Plants<br>• How to Stay Healthy<br>• Machines Then and Now<br>• Why We Recycle | • All About Desert Life<br>• All About Ocean Life<br>• Animals at Night<br>• Incredible Earth | • Animals in Art<br>• Wonders of the Past |
| **5** — 900 headwords | • Materials to Products<br>• Medicine Then and Now<br>• Transportation Then and Now<br>• Wild Weather | • All About Islands<br>• Animal Life Cycles<br>• Exploring Our World<br>• Great Migrations | • Homes Around the World<br>• Our World in Art |
| **6** — 1,050 headwords | • Cells and Microbes<br>• Clothes Then and Now<br>• Incredible Energy<br>• Your Amazing Body | • All About Space<br>• Caring for Our Planet<br>• Earth Then and Now<br>• Wonderful Ecosystems | • Food Around the World<br>• Helping Around the World |

| | Metric measurement | Customary measurement |
|---|---|---|
| Page 8 | 1 liter | 2 US pints |

## Oxford Read and Discover

# Plastic

**Louise Spilsbury**

Read and discover all about Plastic ...

- How do we make plastic?
- Can we recycle plastic?

Word count for this reader: 798 words

Read and discover more about the world! This series of non-fiction readers provides interesting and educational content, with activities and project work.

### Series Editor: Hazel Geatches

 **Level 1** 300 headwords

 **Level 2** ✓ 450 headwords

 **Level 3** 600 headwords

 **Level 4** 750 headwords

 **Level 5** 900 headwords

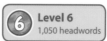 **Level 6** 1,050 headwords

**Also available:**
- 🔊 Audio Pack
- 📖 Activity Book
- 📓 e-Book

You can also enjoy this fiction book from **Oxford Read and Imagine.**

Cover photograph: Alamy Images (Paperclips/Frank Chmura)

SHAPING learning TOGE

**OXFORD**
UNIVERSITY PRESS

www.oup.com/elt

| CEFR |
| --- |
| B1 |
| A2 |
| A1 |

ISBN 978-0-19-46468

9 780194 646888

KT-577-169